Poems by Various (human & non-human) Poets

SHANE NEELEY

AI Art - Poetry
Copyright © 2021 by Shane Neeley

All rights reserved. This book or any portion thereof may not be reproduced or used in any manner whatsoever without the express written permission of the publisher except for the use of brief quotations in a book review.

Special discounts are available on quantity purchases by corporations, associations, and others. Contact details are on the publisher website.

Published by Fort Rock Media
www.FortRockMedia.com

Library of Congress
Control Number: **2020924482**

ISBN **978-1-7362669-2-2**

Printed in the
United States of America

First Printing, 2021

First Edition

Book Design
Melvyn Paulino
www.melvynpaulino.com

Contents

Introduction	*07*
About the Artists	*09*
AI Art & Poems	*13-189*
Tech Specs	*190*
Gift Shop	*191*
Also by this Author	*192*
About the Author	*193*

INTRO-
DUC-
TION

I am no artist, no photographer, no poet; I am an engineer. It floored me when I stumbled upon the digital art technique of neural style transfer. The possibility that I can use my coding skills to make art opened a whole new world for me. In high school I took a computer lab elective of Adobe Photoshop. I stayed hours after class enthralled with digital art. I was mostly cropping out animal heads onto people and vice versa. I loved it; I love hybridizing.

Style transfer images are made with a neural network that combines its own understanding of each image into a new one. When you let this run over and over, it occasionally results in a spectacular blend. The stock images I chose for style transfer mostly relate to nature and technology. For example, gorillas styled with a data center's structured cables, early human stone tools with code screenshots, Native American petroglyphs with iPhone software, etc.

Hybridization is everything: the crossroads of creativity, the reason for the tangled tree of life. It is code merging and company acquisitions. In primatological terms, when there are less than a couple million years of species divergence, interbreeding can happen. This explains how Homo sapiens are a smear of relationships with the other Homo: denisova, neanderthalensis, habilis, and more.

Life mixes it up! Complex life itself started with promiscuity between the ancestors of eukaryotes and bacteria. Even crazier, up to eight percent of the human genome is of retroviral origin, including the functions of mammalian placenta. Without being part virus, we might still lay eggs.

This book mixes it up! People and artificial intelligence wrote the poems. Sometimes both collaborated on a single poem. Amazing friends and professional collaborators made this project possible with their inspiring work. They applied a human eye to the projection of a machine. I trained my poem-generating robot on humanity's poetry greats. It too may inspire you with its layers of calculus and artificial neurons. I hope you enjoy this endeavor in creative data science joined with real human spirits.

Shane
ShaneNeeley.com

On Twitter, Instagram and Facebook as **@chimpsarehungry**

Please sign up for my newsletter to be notified of new books and artwork.

ABOUT THE ARTISTS

I am incredibly thankful for the friends, collaborators, and open-source programmers who made this book possible and our world more poetic.

Humans

Adam Cornford
Adam Cornford is a British-born poet, journalist, and essayist and a **great-great-grandson of Charles Darwin**. From 1987 to 2008, he led the Poetics Program at New College of California in San Francisco. He is the author of four full-length poetry collections: *Shooting Scripts* (1978), *Animations* (1988), *Decision Forest* (1998), and *Lalia* (forthcoming, 1/2021), as well as the book-length documentary poem *Liber Ignis* (2013) in collaboration with fine printer and book artist Peter Koch and several chapbooks.

I. Wimana. C
I. Wimana. C is among other things: a music enthusiast, songwriter, poet, author, and screenwriter, with 10+ years' experience working in the creative world. From his home island of Trinidad and Tobago, he has bridged the gap between local and International artistry, writing for clients on a global scale. His words are his passion and are rooted in life itself. Author of *ECLIPSE* (2020) and *The Puzzle: Finding That Missing Peace* (2016).

Roman Veretelnik
Roman Veretelnik captures the eye with his simple yet direct tones of lyrical writing. Inspired by the hope of peace that comes from above, Roman, a Ukrainian-American, grew up in Oregon and enjoys helping others. Through his inspirational writing, he will leave you with lingering thoughts of wonder and awe.

Brad C. Robertson
Brad C. Robertson is an American poet from South Carolina. He's regarded as a relaxing person who loves to experience and write about the beauty of nature. Despite this, he's also penned a number of startlingly dark poems about the hardships of everyday life, so it can be hard to pin down his stylistic preferences. Author of *50 Poems and the Kitchen Sink* (2019) and *The Art of Heartbreak* (2019).

T.M. Foxglove
T.M. Foxglove, a seasonal poet, inspired by the inhospitable autumn and winters of the Pacific Northwest. Intrigued by the withering leaves and blackened silhouettes of deciduous trees that illuminate the earth with old and new life. A mother of daughters and world traveler, T.M. Foxglove takes you through her written artistry of life, death, and creation of the spiritual and natural world.

Non-Humans

Poem-Writing-Robot

Transformer model trained on an 800,000 line corpus of publicly available poetry. It is blind; it can't see the image, so a human poet described the image in the title or first line of the poem (in **bold**), then the robot did the rest.

Photo-Art-Robot

Neural style transfer model built with PyTorch and ran on Amazon Web Services hardware. Trained to combine technological images with nature scenery styles and vice versa. Shane Neeley curated the images. There is more information about this process in the back of the book.

AI ART & POEMS

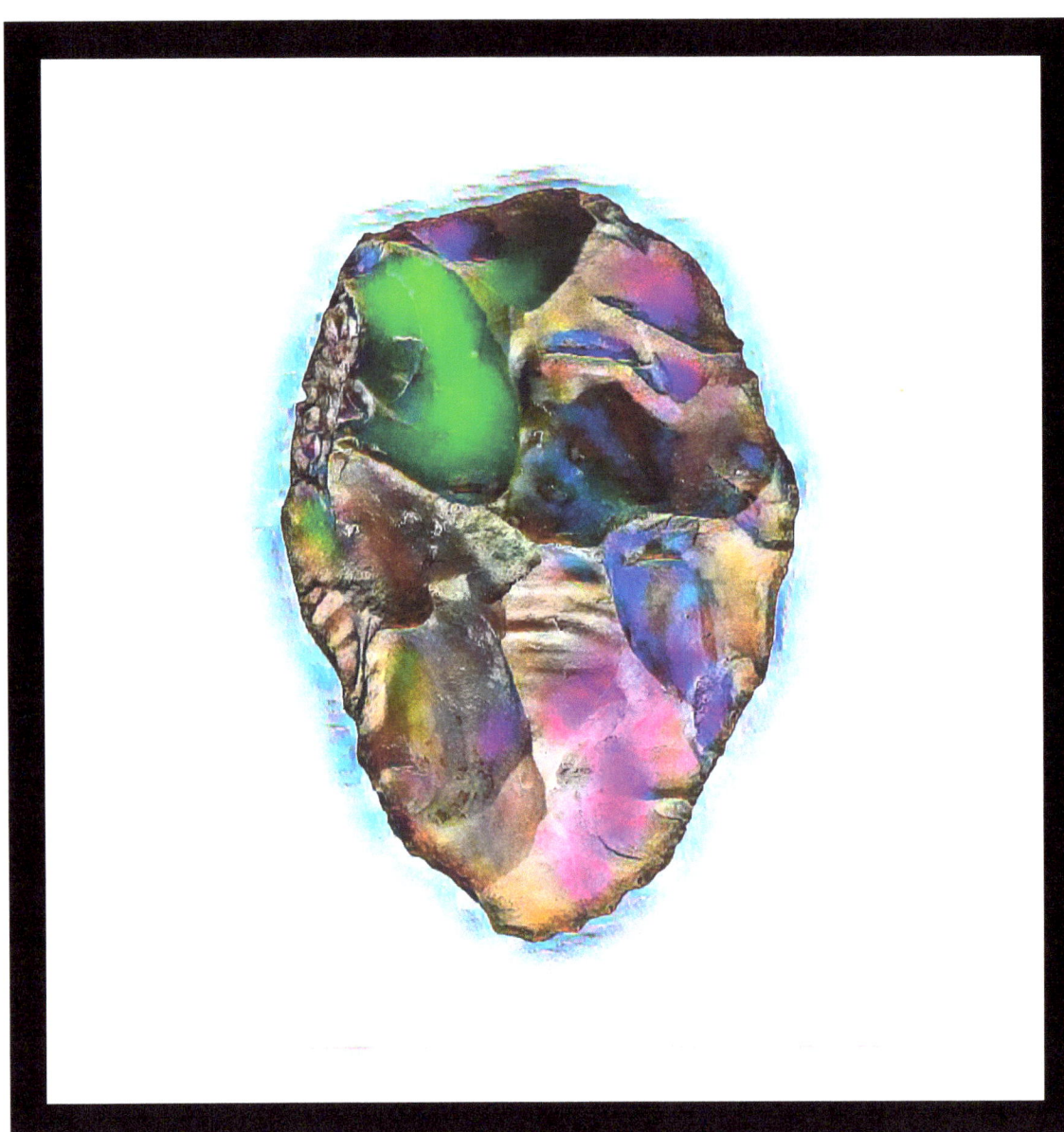

AI-Ku 1

Obsidian blade
Chipped out of code by code tools
Glints virtual light

—Adam Cornford

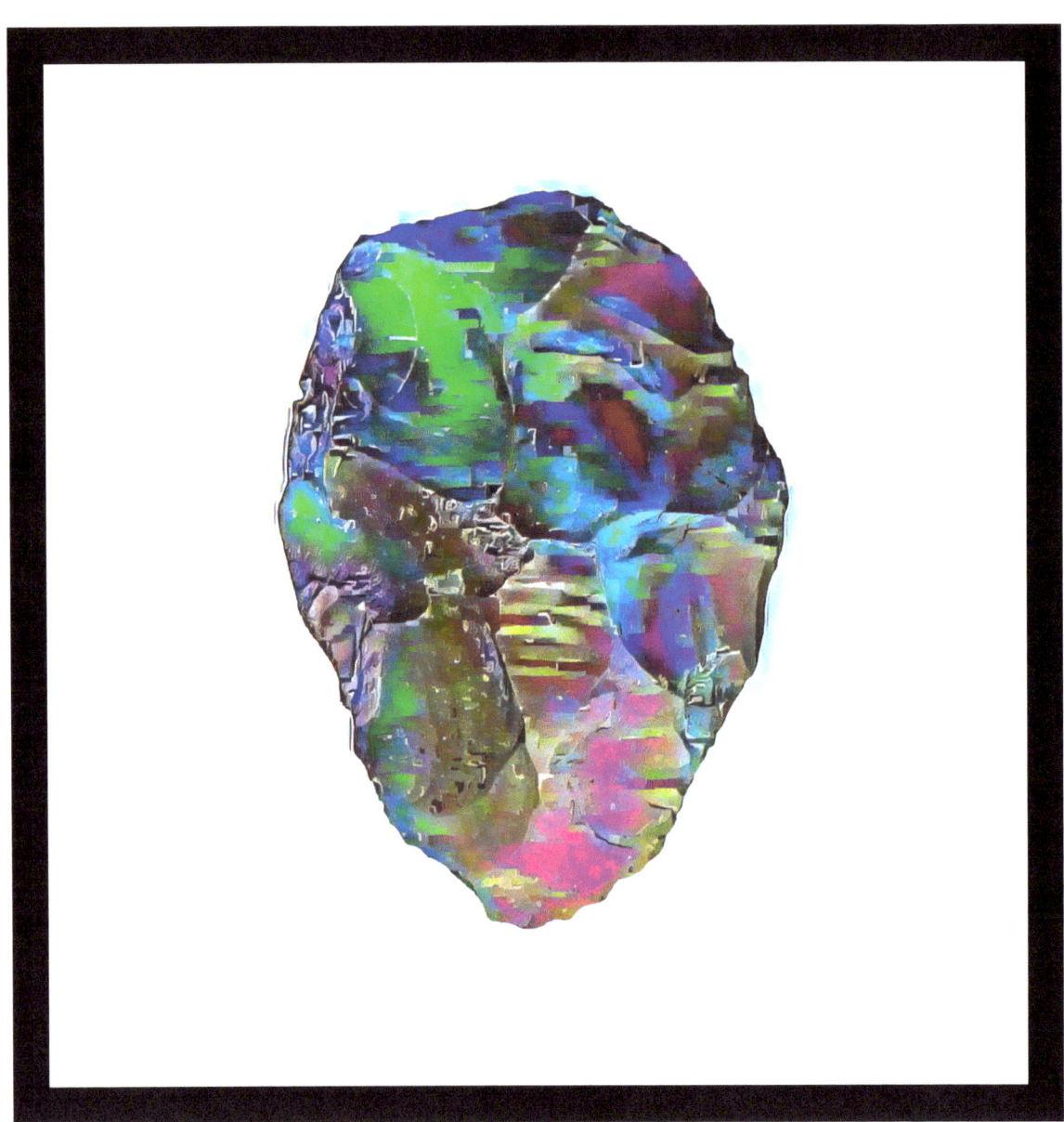

AI-Ku 2

Obsidian blade
Facets, hues illuminated
By machine learning

—Adam Cornford

AI-Ku 3

Obsidian blade
Grown into a black stone mask
Its maker eyeless

—Adam Cornford

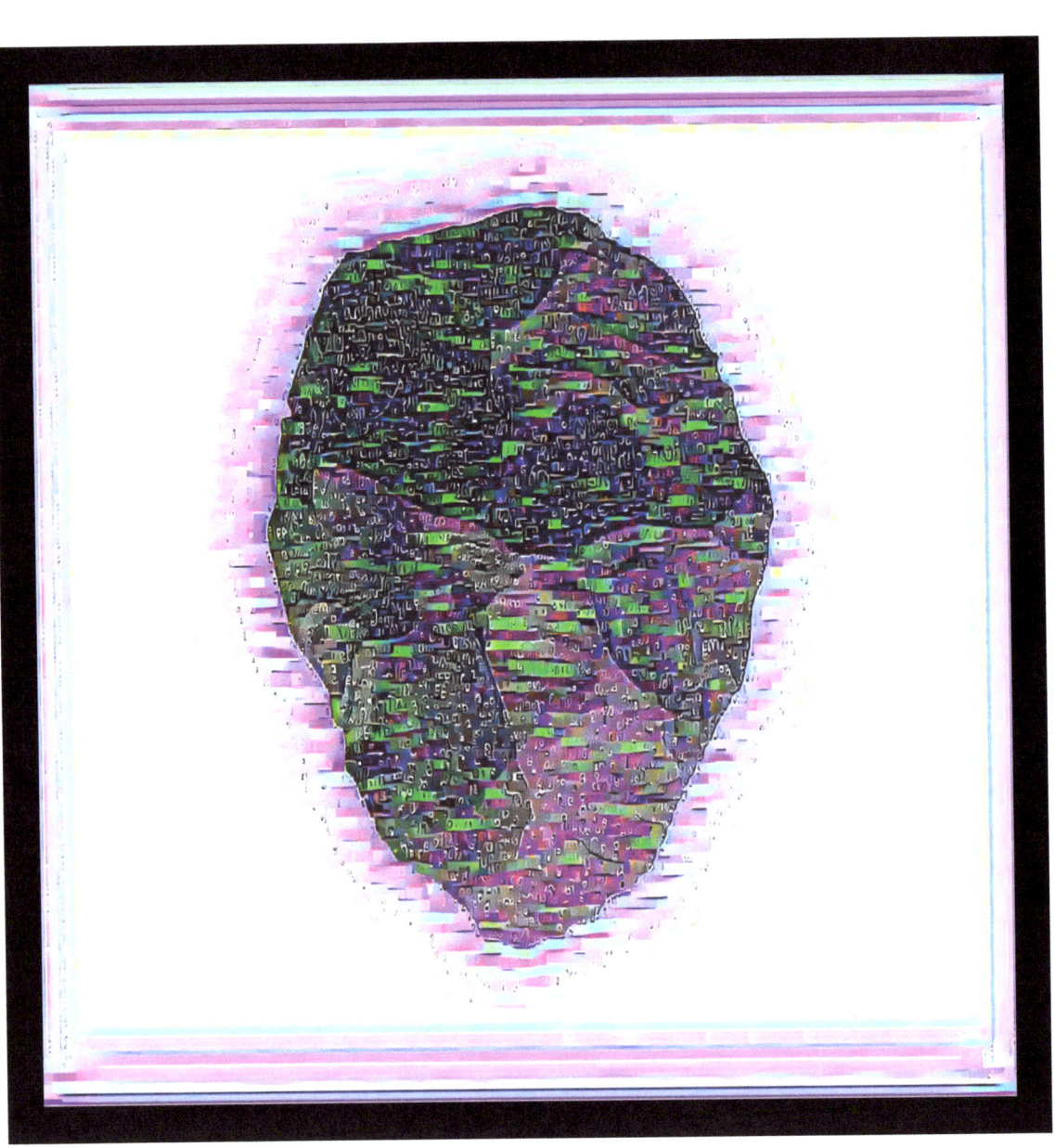

AI-Ku 4

Obsidian blade
Becomes an encrypted map
Our first continent

—Adam Cornford

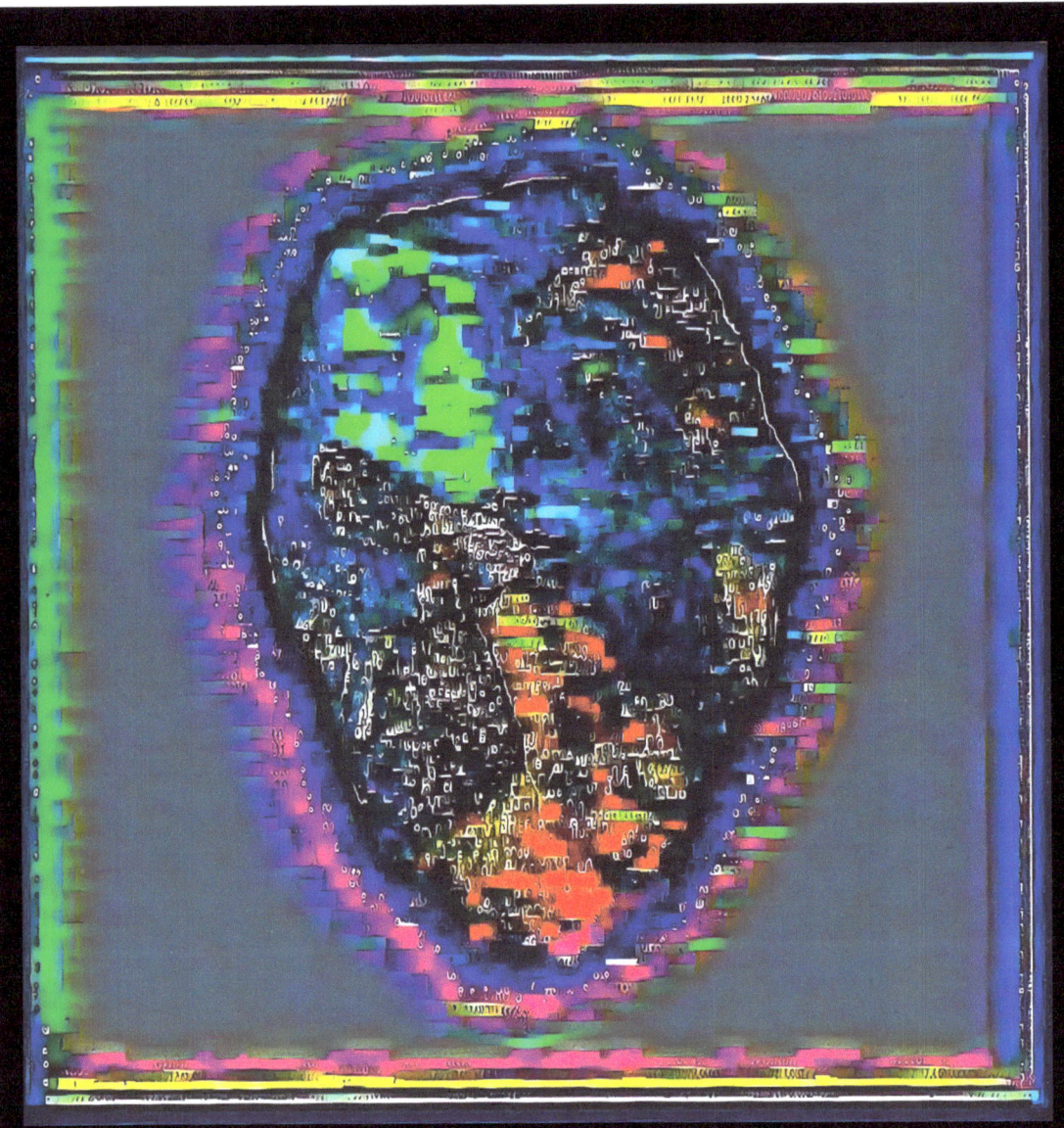

AI-Ku 5

Obsidian blade
Dissolves into pothook script
From after writing

—ADAM CORNFORD

Tree-Dweller

I see a monkey on the limb
And he sees me, look at him!
Cocky monkey in your tree,
Go right ahead, stare at me!

—BRAD C. ROBERTSON

Introspective Trochilidae

A single thought unknown
Rigid in notion by the enigma of the world
Taking flight below the cosmos

—T.M. Foxglove

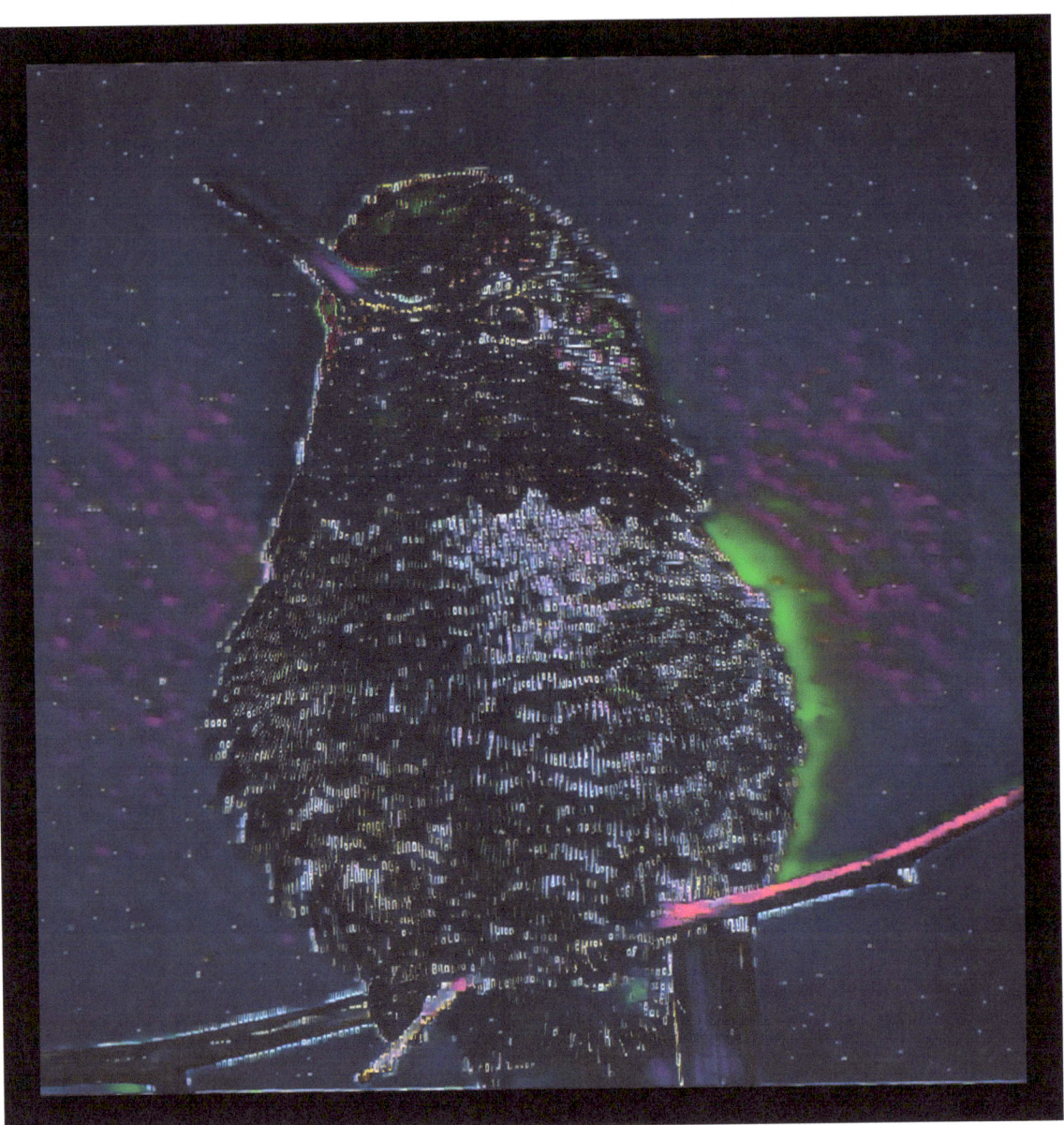

Backlit Bird

Ruffled feathers, eerie glow,
What secrets does the birdie know?
He quietly refuses to reveal
Whatever mystery he conceals.

—Brad C. Robertson

Ruffled pixels perched on a limb awaiting flight

Soon, like a silver bullet, ricochets in the throat.
But they were not the first
There was a brief commercial in Spanish
Then there was no sun

—Poem-Writing-Robot

Wiggling Wall

Wheresoever these hands touch,
It surely will be far too much.
Can sanity and strength survive
Within the fingers as they writhe?

—Brad C. Robertson

Desperate hands reach unto the skies,
Forsaken in darkness, trapped in recluse,
A world void of love will meet demise,
If salvation isn't the path we choose.

—I. Wimana. C

Shades of wonder hindered by the indestructible barricade
and my daughter says my heart has sunk.
The air has blanched into a calm here, says God.
But this didn't make them rich.

—Poem-Writing-Robot

Impassable Firn

Frozen flood, engulfed by mystery
Holding truth to stories flowing through time
Darkened beneath the ground by a motionless rush
Enveloping the earth with wonder and might.

—T.M. Foxglove

Puzzle-Piece Landscape

The jigsaw puzzle of the horizon
As it fills itself in, falling into place
Where twilit sky meets darker ground
In this anomaly of time and space.

—Brad C. Robertson

Some evaporate, some rise up.
Some are dim, some are bright.
Some flow just to dry on rocks,
some flow and water a flock.

—Roman Veretelnik

He who didn't cause it,
Expressed and showed care,
He who caused it,
Expressed shame,
One is the problem,
One is the solution.

—Roman Veretelnik

Where love lives, friendships bloom,
The song of happiness sings its tune,
Connections built on unity's laws,
A noble path with a humble cause,
Seeking one thing for all who exists,
A life of peace, love, and bliss.

—I. Wimana. C

A jeer of of absurdity unimpressed by the elder

"Kee-hee!"
We go slow, then go berserking
For it is not for lack
Why shouldst thou not accept it?

—Poem-Writing-Robot

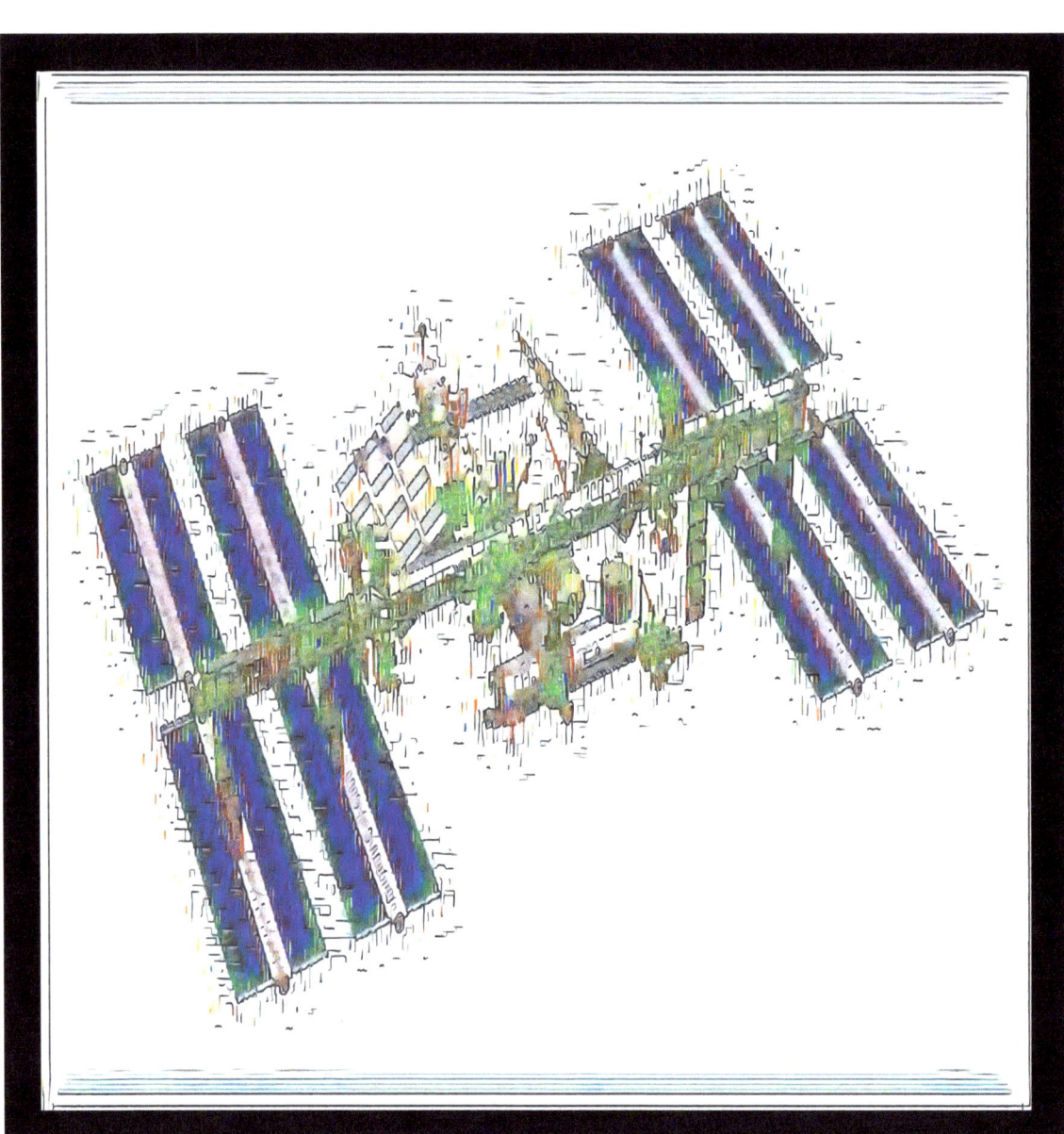

Intellectual beings blessed with the genius of knowledge.
Embraced by the unknown and trusted by mother nature.
Strangers inhabiting a land that granted access to its home,
but our thirst for discovery is as a double-edged sword.
Pouring riches into finding what may never exist,
while our own kind suffers from the injustice of inequality's hatred.

—I. Wimana. C

> **Lustrous copper drifting through the cosmos**
> A few birdsong here, some banter
> This coat was worn in a shoot-in.
> She asked Is the war over yet ?
> War and its men.
>
> — POEM-WRITING-ROBOT

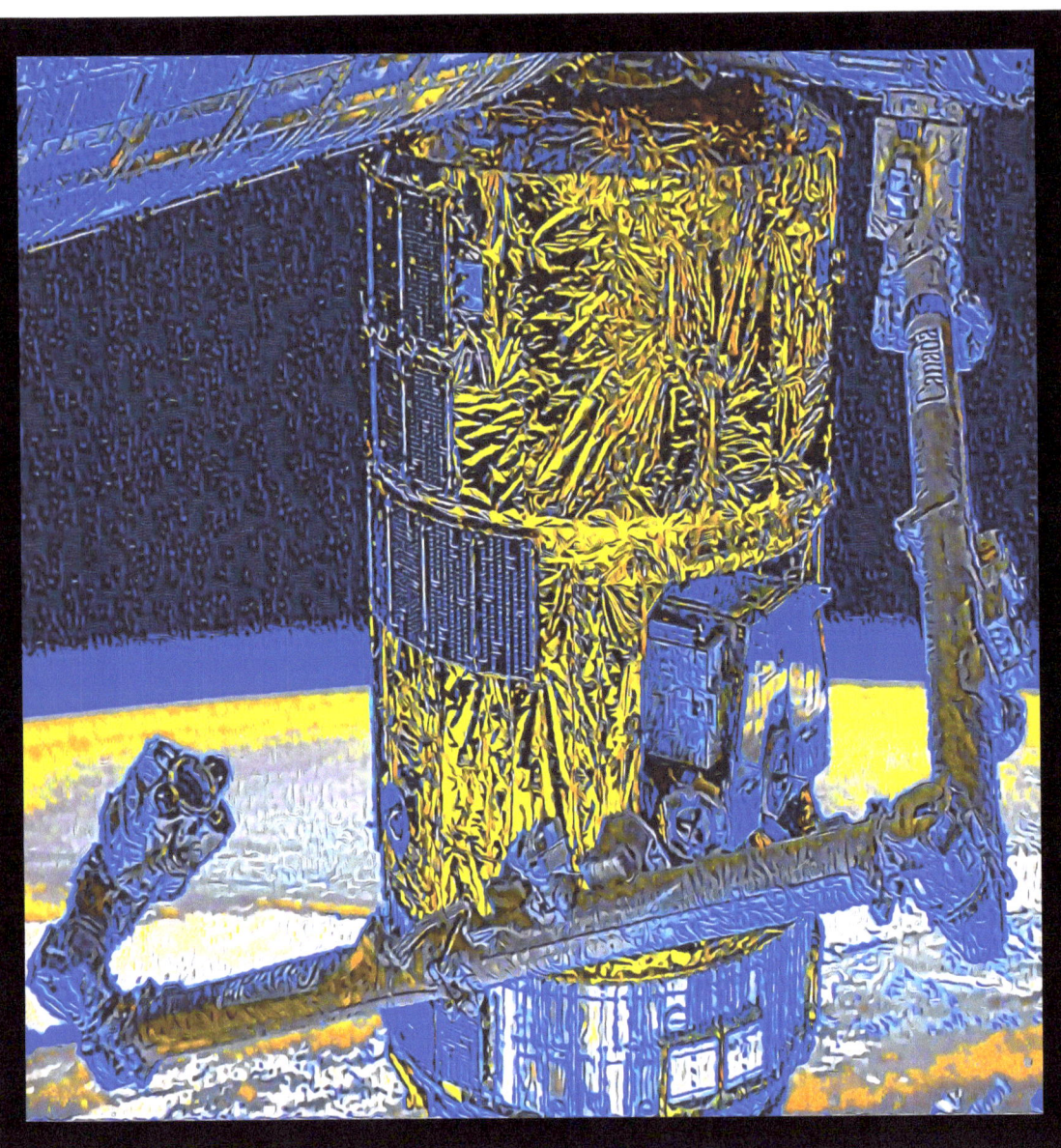

Gliding above a planet entirely known to the universe
Children go missing?
I don't know why.
It is a black day and a black night.
On one breast at a time
All my life

—Poem-Writing-Robot

A hole cannot be filled,
Like an ocean cannot quiet,
Many holes cannot be filled,
Like an agreement in a riot.

—Roman Veretelnik

Contemplating the theory of the beginning

Nobody wanted to hear about us.
And I am a Princess.
I was once very brave.
The corn is a metaphor for love. But never a mule.
The light hummed
And then stopped

—Poem-Writing-Robot

Peace is fleeting in a world on the path to destruction.
Hypnotized by the lure of money and materialism.
Too blinded by sight to listen to the cries of our hearts.
Sometimes solace in solitude, is a necessary precaution,
to reflect on where we are, to change what we've become.

—I. Wimana. C

Depictions of early existence covering the ground with craft

I love the woman there.
The day is bright,
The sea is full of whales
And the moon-light;
The night is full of you.

—Poem-Writing-Robot

AI Art—Poetry   Shane Neeley

What many see as lowly, some see as a blessing.
Wealth is wished for by many, but often attained by few.
Luckily, mother nature has pledged to feed her children.
Upon her land she grants shelter, and from her trees we
shall find food.

—I. Wimana. C

A bounty of golden yield ready to pivot down the bronzed path
Oh, you're a scholar,
Little brother.
What is Hermes doing?
What can you say
Other kids laughing at him
My hair hangs around my waist

—Poem-Writing-Robot

A Quintessential Yearning

Veiled in sorrow
Quenching the eye with sweetness so pure
Collapsing over the earth with an abundance of hope
With a nod to the most dedicated heart.

—T.M. Foxglove

Sinking hearts hanging by the stem of creation
in love with you there is always disagreement.
Fool, vain! Stop
It wasn't designed to be a children's play.
It was designed with you in mind
Children are a garden

—Poem-Writing-Robot

Oh what monsters we've molded within, casting our humanity,
Chasing in the shadows of money, greed, and fame,
Ancestors stare in pain to see a lost generation,
Crying from carved mementos hoping we drift from insanity.

—I. Wimana. C

The original artistry of the earliest souls treading the earth
Indeed, today, I like to joke
I was wrong
It gives the horse a stable and two blankets.
And one brave heart, alas!
It's too much to ask your car to understand you.

—Poem-Writing-Robot

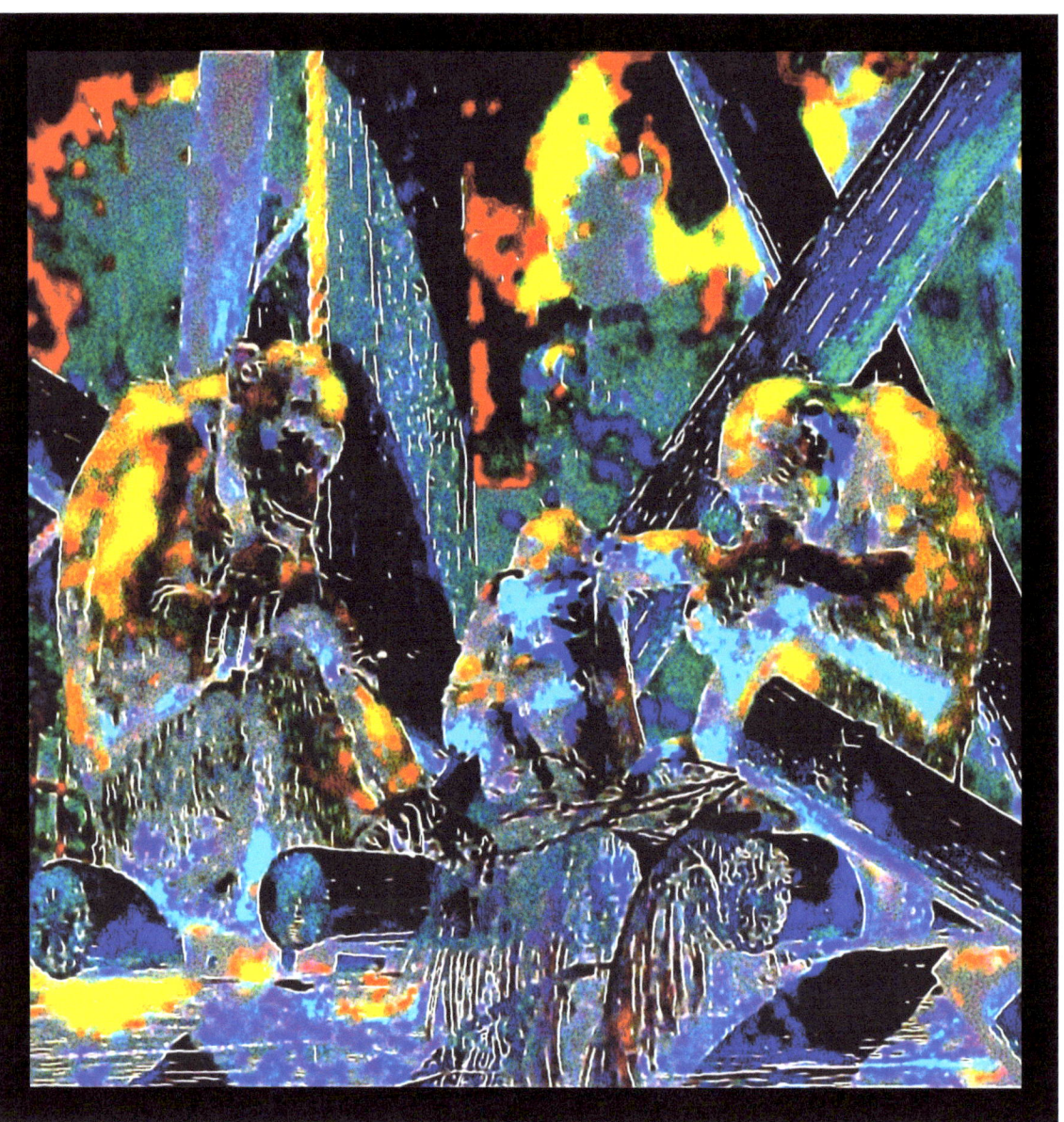

Tinted by the vitality of the living world
the world was endless
And endless in its desperate desire to be endless.
All the political consultants drinking whiskey keep
coffee every Sunday.
How can I know this, but I can.

—Poem-Writing-Robot

Disquieted of sustenance

Bottomless years.
They are not gods:
Why have they come here? Disappearing
You will not be changed.
The next few minutes will be irrevocable
I miss the corn.

—Poem-Writing-Robot

A Comparative Type

Embellished in knowledge, reflective in thought
One cannot know discernment of this prolific mind
Amused by capacity of sentiment
A kindred spirit of our own existence.

—T.M. Foxglove

Irremovable mossy ape

But what were your thoughts then?
I am not trying to look good.
And the sound of rain. O no,
O man!
I wanted to eat grass.

— POEM-WRITING-ROBOT

Rush of water sweeping through the canyon

Nobody's cars are driven through the cold water.
Water-running preps,
The paucity.
A scientist makes the impossible rise from the mere.
The first inspired man.

—Poem-Writing-Robot

AI Art—Poetry   Shane Neeley

Enchanted lands surrounded by verdant trees and flowing waters

No birdsong here.
Stepping crow.
What's your story?
I just want to get the hell out of here
and never get the chance.

—Poem-Writing-Robot

Torched by twilight

Holy the groaning saxophone!
But she knows she is making out in the dark.
The hour came up, and the battle began.
In as the old saying is still the book.

—Poem-Writing-Robot

Gleaming ceaselessly through the night

Indeed I do.
As long as they can, and this is how it ends.
He said, pompously,
Behind me the light glowed.
In my own skin.

—Poem-Writing-Robot

An established horizon illuminating the waves of the sky

Nobody knew what to do.
A lot of time.
Haze streaks all the windows.
They're so nice to work with.
They all fit together.
To keep safe.

— Poem-Writing-Robot

An indistinct city of luminosity

Nobody could connect me ...
I have a ceremonial saber
The traitor came closer.
Then, O let us give an inch.
He breathes and breathes and breathes

—Poem-Writing-Robot

The Marigolds of Machine Learning

Hung on,
A German officer too,
a single sound.
So he carried them.
Compact in your little hand.
In the gutter, sounds of thunder

—Poem-Writing-Robot

What is valued, comes from within
To neglect it, is to live life thin
Who knows the inner depths of man?
Who can fill the void?
Many have, and many do
As so can you

—Roman Veretelnik

Floral Delight

Dew on the grass
Golden drops glistening
Enchanted like the cosmos
Earth twinkles in delight
In hopes of refreshing the souls.

—T.M. Foxglove

She's a flower blooming under the heavens glow, dancing in the wind as her petals hypnotize any entranced by her gaze. Always aiming towards the sun, with eyes unflinching from her truth. Her beauty is rooted in love, ever growing through all trying to break her.

—I. Wimana. C

Silverback male toxicity, protective attitude in his eyes

Where are they going?
Red with iron, white with gold
and a blue hole in their throat.
Yellow with iron.
When I came outside, nobody was allowed to see me.

—Poem-Writing-Robot

Fat neoclassical dyonysian followers

The third snorts of the stereo jazz,
And yet when I touch it, I stop,
I am more like a moth, though
I take the form of an ordinary person

—Poem-Writing-Robot

The witch looms

I was pretty much Indian
The world works best when it's not a fight
it is in any case.
her time had come.
And, O my Soul, do lay thy bones

— POEM-WRITING-ROBOT

Maybe today, my experiment will not fail to soporize

The air here hushed low, low,
And then again: the air.
Aye, he says,
He offers meatloaf sandwich
with mushrooms

—Poem-Writing-Robot

The scientists are fighting

The optics are bad. Nobody Applause.
I told them I wanted closure.
And they were steadfast figures in the swift ocean.
The aftermath of the accident or misunderstanding.
Houses of the mind.

—POEM-WRITING-ROBOT

Labels, Dust, Chores and Opportunity

I am not writing conference papers.
I am writing conference papers.
Look at Herodotus!
when I am not writing memoirs. I didn't think.
My advice: Go to bed.

—POEM-WRITING-ROBOT

Disgraced professor knows his time is due

Nor can I sleep beside their poverty-stratagems.
A neighbor aims in my eyes, spits.
A neighbor aims in the red circle, spits.
You got high expectations,
You got a high school education, you read more than I did.

—Poem-Writing-Robot

What's on the other side of money and measurements?

What pleasure
A big man, big man will take his bundle
And leave the litter?
What!
Do not banish me for digressions.

—Poem-Writing-Robot

Unanswered

Quiet white laboratory
Reveal to me with your tricks
The strange secrets I seek.
What makes us all tick?

—Brad C. Robertson

We partied like this once

Nobody knows when, how and why.
The day is huge and beautiful.
The one promised a lifetime.
The present tense is tense.
It is an accident, I know.
To begin counting

—Poem-Writing-Robot

Lonely alien world, you make no sense to me

The light,
Sleeping Leg mountain, eating mud.
That's all I know,
except maybe this one.
We will see warmth from an angel.

—Poem-Writing-Robot

A ridge connects me to your blue anger

Beside a pillar of light.
We are made so that we love each other.
"Havana."
The child will lie down with the father.
I am an infant before birth.
I was.

—Poem-Writing-Robot

Can we blame her for rebelling? Polluted and abused,
stripped of her serenity, trampling her sacred ground.
She slept well before we came along,
Requesting a place to live and belong.
Her kindness was her demise, trusting gentle
eyes that resort to lies to gain access to her treasures.

—I. Wimana. C

Red Flow

A red that dances sluggishly,
Blue hiding in her cracks.
Flow, flow, red, here.
Be hot, be fluid, be death.

—Brad C. Robertson

I love how you think my dear

Of ants. Hissing?
I thought about ants.
O fond dove!
And so I'll begin again
Again.

—Poem-Writing-Robot

Igniting Life

Incensed with fury, a glow so formidable
Carving a path to make an entrance from beneath the earth
Dismantling all creation for its fiery road
Making fertile the lands to sow.

—T.M. Foxglove

Glowing specks at dusk

We talked on the grass, sniped back to my car.
Not enough mention of us
The day that hung so long
Was the grace of God: to cure Malaria.

—Poem-Writing-Robot

Children will rise up and lead the procession through the dusk

The moon is actually a magnetic field.
"That shall we carry it to the grave,"
quoth the elder, my brother.

—Poem-Writing-Robot

The stillness of azure

Nobody seemed to see.
The last straw was a blow-dryer.
It will be 300 soon.
How?

—Poem-Writing-Robot

A wondrous calligraphy covers the bay

You are the Listener and Reader.
A lamentation to restore what has been.
I am not a river,
I am an ampersand.
Trip tankers waited at the gates

—Poem-Writing-Robot

AI Art—Poetry   Shane Neeley

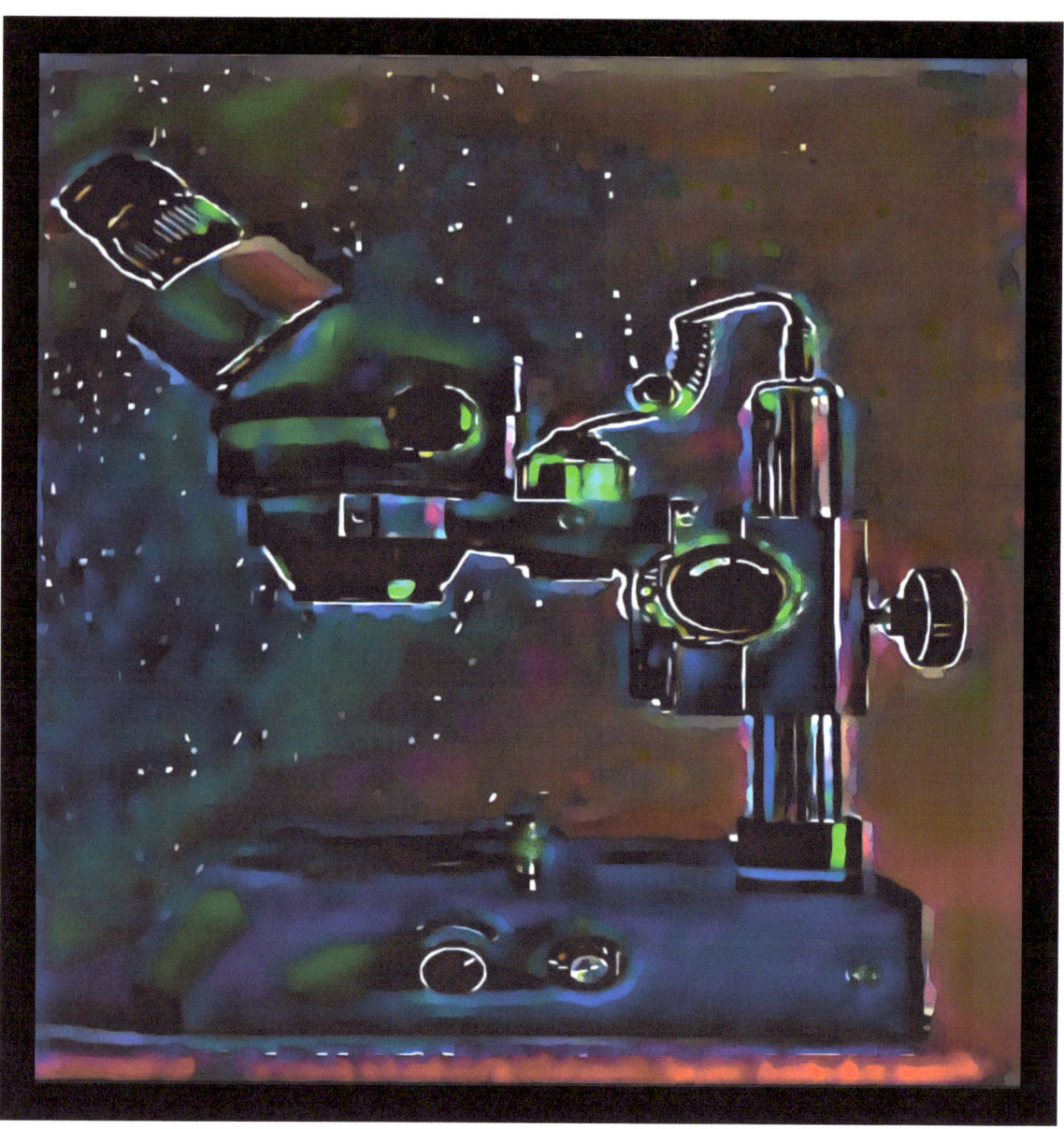

Miniscule life

Wide eyes looking down
Movements ever so delicate
Steadily trudging the glass
Captivating the mind in awe.

—T.M. Foxglove

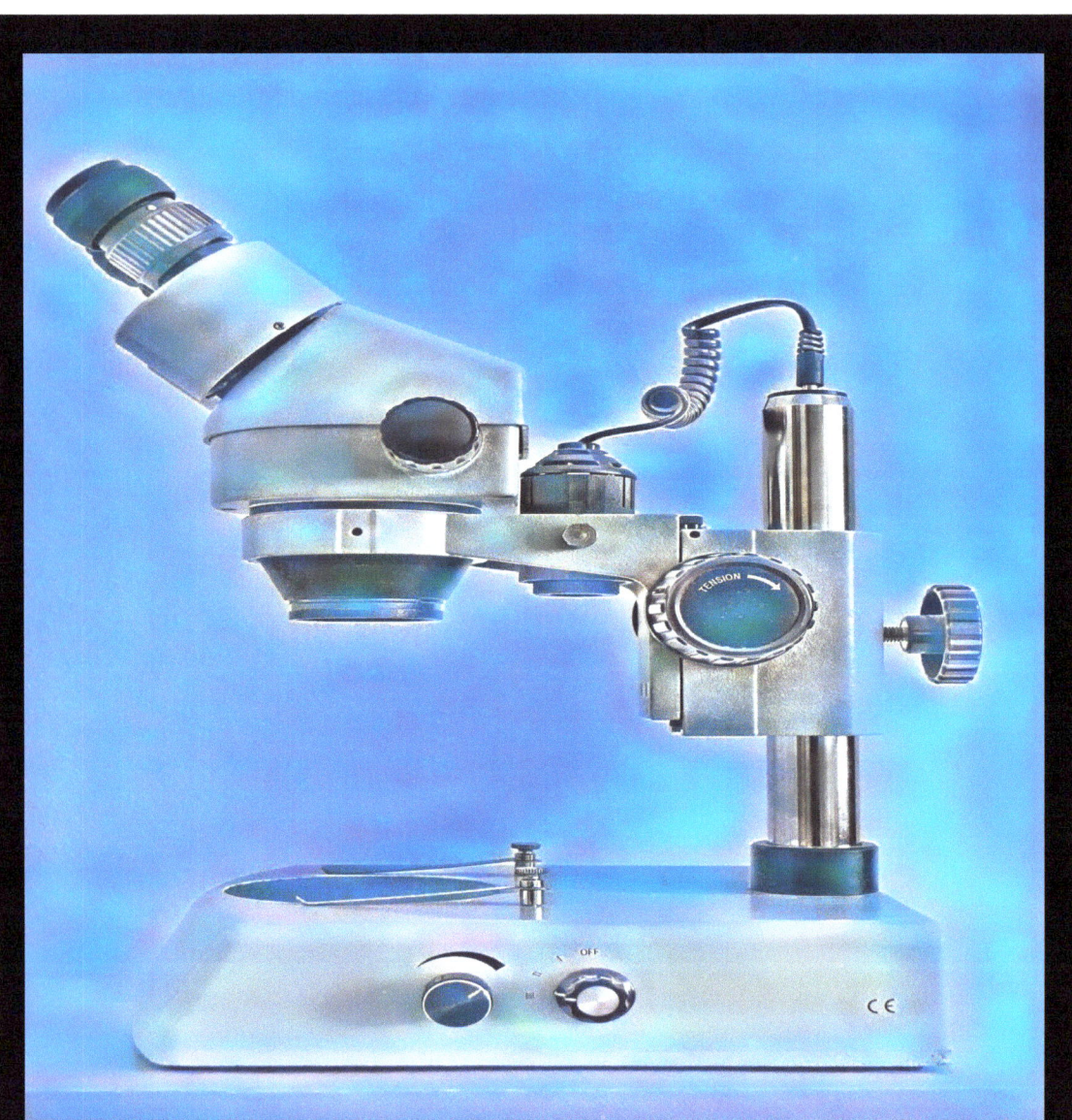

The Discovery of Toothpaste

My teeth dissolve into a sneer of ice
The tips of noses and foreheads turn green for it.
There were always herbs and shrubs in the Roman Bath
Syrup from sibilants' groofs and borracho

—Poem-Writing-Robot

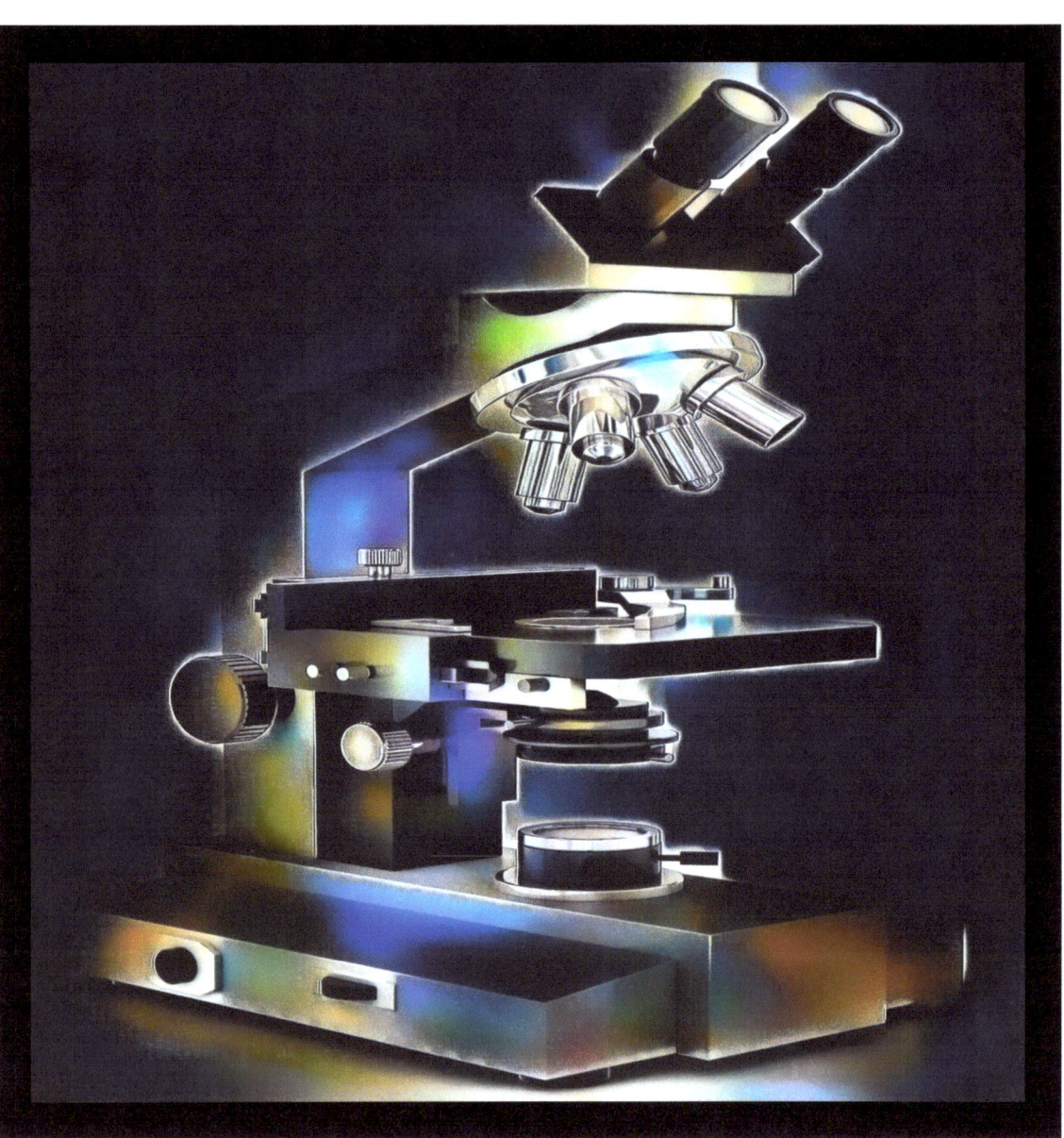

The disease that will bring humanity to its knees,
Can't be seen with the technology of man,
For what will consume us is selfishness and greed,
Spreading destruction throughout our land.

—I. Wimana. C

Ninety-six well microplate
Concentrations, concentrate
A million cells, which one will tell
Which well will illuminate
A cure hiding in the monkey knee
The goal is immortality

—Shane

The Berry Picker's Legacy

The Unlikelies were incorruptible.
It happened it on the field, and off the front porch slamming,
It all happened Right.
He put the jelly down, put the jelly in the cold.
Then he went trippy and calm

—Poem-Writing-Robot

Often times I wish to be,
Just like a waterfall flowing free,
Fluent through life's uncertain path,
With a pure soul and loving heart.

—I. Wimana. C

The first clowns on Mars

Holy! Invertical clouds,
machines nested.
"Ah!" A man cannot be a man but to help man.
To stop the bleeding blood, gently beg the nub.
To no avail. To no avail. In April, no avail.

—Poem-Writing-Robot

Awakened Minds

Human and beast unilluminated
Mysterious to the soul
Restless figures
Leaping in laughter.

—T.M. Foxglove

Salmonberries

The river ran past midnight in the mountains.
When they wake,
Swinging her round body close to my throat, sweetly sifted by the wind.
I've had a month of mostly mental paranoia.
Tender, though I love it.

—Poem-Writing-Robot

Quiet Swimmers

Twinkling droplets bounce against scales
And fins flap like silvered wings
Within the drinking water
By my feet.

—Brad C. Robertson

Fiber-optic eruption

Neon volcano spewing
Up coral high-beams
Into dreamscape skies
Or so it seems.

—Brad C. Robertson

I lay beneath the stars, admiring its glow,
The twinkle that shines upon lonely souls below,
Sharing a bit of warmth to try and ease the pain,
My haven of happiness that makes me smile again.

—I. Wimana. C

Surge of Wind

Awakened by the wind
Stirring through the canyon
Ever so grand
The clouds smile
Watching the spirit guide.

—T.M. Foxglove

We're all a little abstract, flawed with insecurities
but our vulnerability is what makes us beautiful.
A splash in the picture of life, each finding our place
along its journey, cherished puzzles to the whole.

—I. Wimana. C

Suffocating Sky

Again the green cloud.
From whence does it come?
Who will breathe tomorrow
And where goes my sun?

—Brad C. Robertson

A beacon from the earth's heart ascends,
A plea for all the hatred to end,
So unity can once again reign,
Ending the cycle of bloodshed and pain.

—I. Wimana. C

Beneath the stems

Verdant lights of a little world
Where all is large, surreal.
Sunshine softly washing through
As stems sway and softly twirl.

—Brad C. Robertson

A horse that courts the park's ruts
The big frogs croak and circle.
Hesitant at noon, on a log.
What do I seek?
Well-being

—Poem-Writing-Robot

Impatience

Grow faster amidst the fiery heat
Of a scorching summer day.
Rise up, my crops, like a flame
That boldly lights the way.

Brad C. Robertson

Cracks of blue

Splash of color, mostly blue.
Cool, soothing simple hue.
A ripple washing over glass,
Or brittle ice, a broken mass.

—Brad C. Robertson

Eventually I could no longer live.
Brick nod, not wither, fading badly into senile grace.
No ear can hear me.
The pain wasn't unbearable.
This was their life,
They are our life.

—Poem-Writing-Robot

Camping made impossible

Empty base, check for rips!
Keep moving.
Cinnamon, ground mace, and allspice berries.
I did the same thing every day.

—Poem-Writing-Robot

Incandescent Rage

Fluttering in fury
Waiting to acquire sweetness
Bloomed in life so radiant
Enlivening a single existence.

—T.M. Foxglove

TECH SPECS

Poem-Writing-Robot

This was done with a GPT-2 transformer model trained on hundreds of thousands of poems. I used a Python package called *simpletransformers* and a *huggingface* pretrained GPT-2 large model. I built the dataset as a combination of several open sources. I processed the data so that it included newline characters with an average of 6 lines per poem. Then adjusted the maximum sequence length for training and language generation.

For the generation step, I ran the model with a prompt that was the first line of a poem written by a real poet. Then the model filled in the next 5-6 lines. Some editing for grammar and structure and readability was needed. The generator wrote many poems and I attempted to choose one of them that fit the image. Some dissenters would not call this art, but cherry-picking. Well, I happen to like picking cherries, and I avoid the ones with worms.

Photo-Art-Robot

Adapted from the tutorial Neural Transfer Using PyTorch. The underlying principle of neural style transfer is to create a new image based on a content image and a style image. The new image should retain the overall look of the original content image, but be stylistically similar to the style image. The model accomplishes this by retaining two loss functions: a content loss and a style loss. You use a neural network that has been pretrained on image data and perform gradient descent on the pixel values of the new image to minimize both content loss and style loss.

The secret sauce for my images is sheer computation power and patience. I adapted the PyTorch style transfer example to do lots of random sampling of hyperparameters to produce new images. This included ranges of the style weights, the orientation of the images, the style image, rotation of style image, desired depth of convolutional layers, the model itself (vgg11 through vgg19), and gradient descent steps. Each run would be assigned random values for each of these and run the content image through them. Hours and hours on a rented GPU was able to create a few gems out of most images that I threw away. I used a combination of free usage and paid usage stock photos for both content and style.

GIFT SHOP

ShaneNeeley.com/store

If you love this art and would like some in your home or on your person, check out my store. Find mugs, shirts, framed prints, stickers, face masks, and bags for sale.

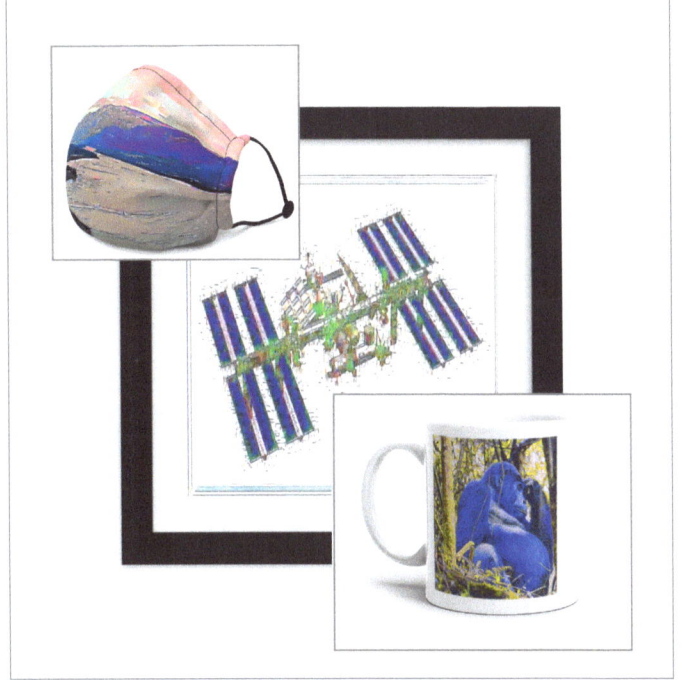

Thanks for reading my poetry book! Use this special coupon code to get a discount on any AI Art in the store: **CHIMPSARESTILLHUNGRY**

If you would like to get in contact with me about any of the art, sign up for my newsletter and reply to the welcome email.

ALSO BY THIS AUTHOR

2021 Release

Stone Age Code illustrates the evolution of improbable data scientists. Though bad at math and often monkey-minded, Shane Neeley became outstanding at coding and building intelligent, creative machines. The universe also underestimated the walking apes; the monkeys weren't supposed to make it to space or conjure AI with clicks on a keyboard. A paleolithic birth of creativity, driven by the same old evolutionary selection pressures, fashioned humankind into artists, poets, comedians, believers, and scientists. Another epic transformation is happening today as people begin to build intelligent machines. This book is a must-read for anyone learning AI, coders, and coders-to-be!

Become an AI forefather to future generations, regardless of your background. In this entertaining approach, you will gain an understanding of deep learning with neural nets, natural language generation, and AI art. Not a technical deep dive—the book relies more on sweeping generalizations and humor for instruction on its subject matter. Containing no equations or code, it still teaches machine learning literacy, and in an amusing way. You can train AI now to boost your imagination—to produce literature, art, comedy, and to tackle challenges at work.

ABOUT THE AUTHOR

Shane Neeley lives in Oregon, writes code, and enjoys taking his daughters on explorations in the Pacific Northwest. A former lab scientist turned programmer, he helped to author eleven scientific papers and build a cancer treatment software company. He traded in a lab coat for a laptop, now using machine learning and artificial intelligence for everything from bioinformatics to art. He is particularly interested in AI for language generation and other means of computational creativity.

His YouTube videos on Python coding have hundreds of thousands of views and hundreds of downvotes (apparently some people do not like how he talks). Shane tries to share a transformative message about technology and biology through comedic writing and speaking. He's been described as rarely serious on serious issues; communicating advanced topics with a spectrum of joking.

He studied at the University of Colorado - Boulder in the Molecular, Cellular and Developmental Biology program, and completed graduate school in Bioengineering at Rice University in Houston, Texas.

www.ingramcontent.com/pod-product-compliance
Lightning Source LLC
Chambersburg PA
CBHW041706160426

43209CB00017B/1759